Solar Energy's Stellar Successes

Charting Progress, Illuminating Breakthroughs, And Achieving Sustainable Energy Futures

Zoltan Fairchild

Contents

1 Chapter 1: Introduction to Solar Power

Harnessing the sun's energy, a seemingly limitless resource, offers a compelling solution to our global energy demands. Solar power, at its core, is the conversion of sunlight into usable energy. This conversion happens through two primary methods: photovoltaic (PV) technology, which directly transforms sunlight into electricity, and concentrated solar power (CSP), which uses mirrors or lenses to focus sunlight onto a receiver, generating heat that drives a turbine to produce electricity. Both are crucial components of a sustainable energy future and each presents unique characteristics.

Unlike fossil fuels, solar energy is clean and renewable, producing little to no greenhouse gas emissions during operation. This significantly reduces our carbon footprint, contributing to a healthier planet. However, the initial investment for solar systems can be substantial. Factors such as system size, location, and government incentives influence the overall cost. Fortunately, advancements in technology and increased economies of scale continually reduce the cost of solar energy, making it increasingly accessible and affordable.

Photovoltaic (PV) systems are the most common type of solar energy technology. These systems use solar panels composed of numerous photovoltaic cells. Sunlight striking these cells generates an electric current. The simplicity of PV systems, their scalability from small residential installations to large-scale solar farms, and their ease of integration into existing electrical grids, contribute significantly to their widespread popularity. Many homeowners now opt for rooftop solar systems, reducing their reliance on the electricity grid and lowering their energy bills.

Concentrated Solar Power (CSP) systems offer a different approach. CSP plants use mirrors or lenses to concentrate sunlight onto a receiver, creating high temperatures. This heat is then used to generate steam, which drives a turbine connected to a generator, producing electricity. CSP systems are particularly valuable due to their ability to incorporate thermal energy storage. This means they can continue generating power even after sunset, addressing the intermittency that sometimes plagues solar energy. However, CSP systems generally require larger land areas compared to PV systems.

The choice between PV and CSP depends on several factors. PV systems are often preferred for smaller installations and locations with ample sunlight but limited land. CSP, with its potential for energy storage, may be more suitable for large-scale projects, particularly in areas with high solar irradiance and where continuous power supply is crucial. Regardless of the chosen technology, solar energy is a powerful tool in our efforts to transition to a cleaner and more sustainable energy system. The ongoing advancements in both PV and CSP technologies promise even greater efficiency and cost-effectiveness in the years to come. The widespread adoption of solar power is not merely a technological achievement; it represents a significant step towards environmental responsibility and energy independence. The future of energy is bright, and solar power will play a pivotal role in illuminating that future. Understanding the basics of solar energy is the first step towards embracing its transformative potential.

1.1 What is Solar Energy?

Sunlight, the radiant energy emitted by our sun, is the foundation of solar energy. It's a clean, renewable resource unlike fossil fuels, constantly replenished by a nearly inexhaustible source. This energy, reaching Earth in the form of photons, carries the potential to significantly alter our energy landscape, offering a pathway towards a sustainable future. Understanding this potential is crucial.

Harnessing this power involves two primary methods: photovoltaic (PV) technology and concentrated solar power (CSP). PV systems directly convert sunlight into electricity using semiconductor materials, typically silicon. These materials absorb photons, causing electrons to flow, generating an electric current. Think of it as a natural, ongoing chemical reaction that produces usable energy. The efficiency of this conversion, however, varies based on factors like sunlight intensity, cell temperature, and the specific materials used.

Concentrated solar power, on the other hand, utilizes mirrors or lenses to focus sunlight onto a receiver. This concentrated heat then generates steam, which drives a turbine connected to a generator, creating electricity. CSP systems differ from PV in their approach; they leverage the thermal energy of sunlight, creating electricity through a more traditional power generation process. This approach often incorporates thermal energy storage, allowing for electricity generation even after sunset.

The benefits of solar energy extend beyond its clean energy production. It's a decentralized energy source, reducing reliance on large, centralized power plants and improving energy security. Further, the environmental impact is significantly lower compared to fossil fuels, reducing greenhouse gas emissions and air pollution. While manufacturing solar panels requires resources and energy, their operational emissions are negligible, fostering a greener energy transition.

Despite these advantages, challenges exist. The intermittency of solar energy—its dependence on sunlight—requires effective energy storage solutions and grid integration strategies. Furthermore, the initial cost of solar installations can be substantial, although decreasing steadily. However, various government incentives and financing options are increasingly making solar energy more accessible to individuals and businesses.

The efficiency of solar technology is continually improving. Researchers are exploring new materials and designs for both PV and CSP systems to enhance energy conversion rates and reduce costs. For example, perovskite solar cells show promise due to their potential for high efficiency and low manufacturing cost. Similar advancements are being made in CSP systems, particularly in the development of more efficient thermal storage technologies.

The future of solar energy is bright, literally and figuratively. Its integration into existing grids and development of smart grids are key areas of focus, enabling better management and distribution of solar-generated electricity. Furthermore, advancements in energy storage technology, such as batteries and pumped hydro, are overcoming the intermittency challenge, making solar power a more reliable source of energy.

In conclusion, solar energy represents a powerful tool in our fight against climate change and our quest for a sustainable energy future. It offers a clean, renewable, and increasingly affordable alternative to fossil fuels. While challenges remain, ongoing technological advancements and supportive policies are driving its widespread adoption, positioning solar energy as a key player in the global energy transition. Understanding its various forms, its applications, and the ongoing improvements in efficiency are vital in fully appreciating its potential to reshape our world. The future demands energy innovation, and solar energy stands as a leading example of this necessary and potentially game-changing innovation.

1.2 Types of Solar Technologies

Photovoltaic (PV) systems directly convert sunlight into electricity using semiconductor materials like silicon. These systems are the most common type of solar technology, found everywhere from small rooftop installations to massive solar farms. Their modular design allows for scalability, making them suitable for diverse applications. While silicon-based PV cells dominate the market, research continuously explores alternative materials such as perovskites and cadmium telluride, promising higher efficiencies and lower manufacturing costs. Perovskites, in particular, are attracting significant attention due to their potential for high power conversion efficiency and flexible application. However, long-term stability remains a challenge that researchers are actively addressing to ensure the longevity and reliability of these next-generation solar cells. The widespread adoption of PV technology is fueled by its relatively low environmental impact compared to fossil fuels, although concerns remain regarding the manufacturing process and the eventual disposal of solar panels. Recycling initiatives are gaining traction to mitigate these issues, promoting a more circular economy for solar technology. Understanding the diverse range of PV technologies available, from monocrystalline to thin-film, is crucial for making informed decisions about solar energy investments.

Concentrated Solar Power (CSP) systems, unlike PV, use mirrors or lenses to concentrate sunlight onto a receiver, generating heat that drives a turbine to produce electricity. This approach differs significantly from PV, offering the advantage of thermal energy storage. This means CSP plants can continue generating power even after sunset, addressing the intermittency challenge inherent in solar energy. Several CSP technologies exist, including parabolic troughs, power towers, and dish Stirling systems, each with unique characteristics and efficiencies. Parabolic troughs, for example, utilize long, curved mirrors to focus sunlight onto a pipe containing a heat-transfer fluid, while power towers employ a field of heliostats (sun-tracking mirrors) to concentrate sunlight onto a central receiver atop a tower. The choice of CSP technology depends on factors such as geographic location, climate, and available land area. While CSP offers dispatchable power, its higher initial investment cost compared to PV can be a barrier to entry. However, ongoing technological advancements and decreasing costs are making CSP increasingly competitive. Further advancements in thermal storage technologies are crucial for enhancing the economic viability and operational flexibility of CSP plants.

Hybrid solar systems combine the strengths of both PV and CSP technologies. These systems often incorporate PV panels to generate electricity during peak sunlight hours, supplemented by a CSP component with thermal storage for extending power generation beyond daylight. This hybrid approach optimizes energy production, maximizing the utilization of available solar resources. Such integration offers significant advantages in terms of energy security and grid stability, particularly in regions with fluctuating solar irradiance. Designing these systems requires careful consideration of factors like energy storage capacity, power output requirements, and the overall cost-effectiveness of the integrated solution. They are particularly useful in areas where land is abundant and energy storage is critical for reliable power supply. The complexity of hybrid systems requires specialized expertise in both PV and CSP technologies, and the higher initial cost may be a limiting factor for smaller-scale applications. However, the long-term benefits in terms of enhanced reliability and efficient energy utilization make it a compelling option for large-scale power generation and specific niche applications.

Beyond PV and CSP, emerging solar technologies are constantly evolving. These include organic photovoltaics (OPVs), which utilize organic materials to create flexible and lightweight solar cells, suitable for applications where traditional silicon-based cells might not be practical. Another area of innovation is perovskite solar cells, already mentioned, which exhibit high power conversion efficiencies but still require significant improvements in long-term stability. Research into dye-sensitized solar cells (DSSCs) also continues, promising lower manufacturing costs and potentially higher efficiencies under low-light conditions. Each of these newer technologies presents unique challenges and opportunities, with potential applications ranging from building-integrated photovoltaics to wearable electronics. Their commercial viability largely depends on further advancements in materials science, manufacturing processes, and cost-effective production techniques. Continued investment in research and development is crucial for unlocking the full potential of these emerging solar technologies and diversifying the renewable energy landscape.

2 Chapter 2: Photovoltaic Technology

Sunlight, the most abundant renewable resource, is harnessed by photovoltaic (PV) technology to generate electricity. PV cells, the fundamental building blocks, are crafted from semiconducting materials, typically silicon. When sunlight strikes these cells, photons excite electrons, generating an electrical current. This process, the photovoltaic effect, is the heart of solar energy production. The efficiency of this conversion process, however, is a critical factor determining the overall effectiveness of a solar panel.
Current silicon-based PV cells, while dominant in the market, possess inherent limitations in efficiency. Their efficiency is constrained by factors such as light reflection, electron-hole recombination, and the inherent band gap of silicon. Ongoing research focuses on enhancing these cells' performance through various strategies. These include improvements in surface tex-

turing to reduce reflection, implementation of advanced anti-reflective coatings, and the use of sophisticated doping techniques to optimize electron flow. Developing more efficient silicon cells remains a critical area of ongoing development, driving cost reductions and increasing the energy output.

Beyond silicon, researchers actively explore novel materials with the potential to significantly enhance photovoltaic efficiency. Perovskite solar cells, for example, have emerged as a promising alternative. Their remarkable ability to absorb a broader spectrum of sunlight compared to silicon offers the potential for substantially higher energy conversion efficiency. These materials offer a path toward cheaper and more flexible solar cells, paving the way for applications beyond traditional rigid panels. Challenges remain, however, regarding long-term stability and scalability, issues actively being addressed by researchers worldwide.

Organic photovoltaic (OPV) cells represent another exciting avenue of research. These cells are fabricated using organic polymers or small molecules, allowing for flexible, lightweight, and potentially low-cost solar cells. The inherent flexibility of OPV cells makes them particularly appealing for integrating into various surfaces, including clothing, building materials, and even flexible electronics. While their efficiency is currently lower than silicon-based cells, significant advancements are pushing their performance toward commercial viability. Overcoming limitations in efficiency and longevity are currently being addressed by researchers looking for new materials and designs.

Thin-film solar cells, a class encompassing various materials like cadmium telluride (CdTe) and copper indium gallium selenide (CIGS), offer another approach to improve efficiency and reduce manufacturing costs. These cells are manufactured by depositing thin layers of semiconductor material onto a substrate, resulting in lighter and potentially cheaper cells than conventional silicon-based technologies. While their efficiency is generally lower than that of high-efficiency silicon cells, their cost-effectiveness and potential for large-scale production make them a significant player in the solar energy market. Further improvements in materials and processing techniques continue to refine the cost/performance balance, making them very attractive for large-scale applications.

The development of tandem solar cells, which combine multiple layers of different semiconductor materials, promises a significant leap in efficiency. By stacking cells with different band gaps, these configurations can absorb a wider range of the solar spectrum, leading to higher overall energy conversion rates. This innovative approach has the potential to overcome limitations in single-junction cells. Research is concentrating on creating a stable and efficient tandem structure, with several promising material combinations currently under investigation.

Quantum dot solar cells represent another frontier in photovoltaic technology. These cells utilize semiconductor nanocrystals, or quantum dots, to absorb sunlight. The unique optical properties of quantum dots allow them to efficiently capture photons across a wide range of wavelengths, potentially achieving extremely high efficiencies. The ability to fine-tune the size and composition of quantum dots offers excellent control over their energy absorption characteristics, maximizing energy conversion. While still in their early stages of development, quantum dot solar cells have shown promising results and warrant continued investigation for their future potential.

Improving the efficiency and reducing the cost of photovoltaic technology are crucial for its widespread adoption. Continued advancements in materials science, device engineering, and manufacturing techniques are essential for making solar energy an even more sustainable and cost-effective energy solution. The ongoing research and development in these areas point toward a future where solar energy plays an increasingly dominant role in meeting global energy demands. This future involves not only improved efficiency but also improved durability, reduced manufacturing costs and improved methods of recyclability for environmental sustainability.

2.1 Solar Cell Fundamentals

Sunlight, the seemingly inexhaustible energy source, fuels the photovoltaic (PV) effect, the cornerstone of solar cell technology. This effect, simply put, is the conversion of light energy directly into electrical energy. When photons, or light particles, strike a semiconductor material within a solar cell, they excite electrons, freeing them from their atomic bonds. These freed electrons flow through an external circuit, creating an electric current – the very essence of solar power generation. This fundamental principle underpins the operation of all solar cells, from the simplest designs to the most advanced. Understanding this is crucial for anyone venturing into the field of solar energy.

The basic structure of a solar cell is remarkably simple yet elegant. It typically consists of two layers of semiconducting material, often silicon, with different electrical properties—a p-type layer and an n-type layer. The p-type layer has an excess of “holes” (positive charge carriers), while the n-type layer has an excess of electrons (negative charge carriers). When these layers are brought together, a junction forms, creating an electric field that drives the flow of electrons when sunlight strikes the cell. This seemingly simple junction is the heart of the energy conversion process, a marvel of materials science and physics ingeniously combined. The efficiency of this conversion, however, depends greatly on factors we will explore in detail later.

Different materials can be used to construct solar cells. Silicon, the most widely used material, dominates the market due to its abundance, relatively low cost, and acceptable efficiency. However, researchers are constantly exploring alternative materials to improve efficiency and reduce costs further. These include amorphous silicon, cadmium telluride (CdTe), copper indium gallium selenide (CIGS), and perovskites, each with its own advantages and disadvantages concerning efficiency, cost, and environmental impact. The search for the perfect material continues, driven by the ongoing need to increase the efficiency and affordability of solar energy. This constant evolution is an integral part of the advancements in the technology.

The efficiency of a solar cell—its ability to convert sunlight into electricity—is a critical performance indicator. Various factors influence this efficiency. The band gap, the energy difference between the valence and conduction bands in the semiconductor material, plays a significant role. A narrower band gap means the cell can absorb more of the solar spectrum, but it may also reduce the voltage output. Similarly, the quality of the semiconductor material, the presence of defects and impurities, and the effectiveness of anti-reflective coatings significantly impact the efficiency of the energy conversion. Understanding these subtle interactions helps in designing improved cells.

Beyond the material and its properties, the design and fabrication of the solar cell itself are crucial for maximizing its efficiency. Careful control of the doping process, which introduces impurities into the semiconductor to create the p-type and n-type layers, is essential. The geometry of the cell, the thickness of the layers, and the contact metallization all contribute to the overall performance. Advanced techniques like surface passivation, used to reduce recombination of electrons and holes, help boost efficiency. All of these minute details contribute to optimizing the solar cell's ability to generate electricity efficiently. The entire process is highly sophisticated, a testament to the advanced manufacturing techniques required.

Finally, understanding solar cell fundamentals is not just about the physical principles and materials involved; it's about the broader context of sustainable energy. The choice of materials for solar cells has environmental implications. The manufacturing process consumes energy and resources, generating waste. Analyzing the life cycle of a solar cell—from raw material extraction to disposal—is crucial for assessing its overall environmental impact. This perspective is essential, especially as we integrate solar energy into a broader plan for a more sustainable future. Responsible innovation demands both technological advancements and a clear understanding of the bigger picture. Choosing environmentally friendly options throughout the process is paramount.

2.2 Efficiency Improvements

Improving the efficiency of photovoltaic (PV) cells is paramount to the widespread adoption of solar energy. Current silicon-based cells, while effective, still have room for significant enhancement. Research into novel materials and cell architectures offers promising pathways to achieve higher conversion efficiencies, ultimately reducing the cost of solar electricity. One promising avenue is the exploration of perovskite solar cells. These cells, characterized by their unique crystal structure, exhibit exceptionally high power conversion efficiencies, sometimes exceeding those of traditional silicon cells in laboratory settings. However, challenges remain in terms of long-term stability and scalability for mass production. Addressing these challenges is key to unlocking the full potential of perovskite technology. Continuous refinement of manufacturing processes, along with improved encapsulation techniques, are vital steps towards commercially viable perovskite-based solar panels.

The development of tandem solar cells represents another exciting frontier in efficiency enhancement. These devices combine different semiconductor materials, each absorbing a specific portion of the solar spectrum. By stacking multiple layers with complementary absorption properties, tandem cells can capture a broader range of sunlight, thus boosting overall efficiency significantly. For example, combining a silicon bottom cell with a perovskite top cell can create a highly efficient tandem device, capitalizing on the strengths of both material systems. Careful consideration must be given to optimizing the interface between the different layers, ensuring efficient charge transfer and minimizing energy losses.

Beyond material innovation, advancements in cell architecture also play a crucial role. For instance, researchers are actively exploring the potential of light trapping techniques to increase the amount of sunlight absorbed within the PV cell. This can involve incorporating textured surfaces or nanophotonic structures that scatter light, increasing its path length and the probability of photon absorption. Similarly, improvements in the anti-reflective coatings applied to the cell's surface can further enhance light absorption. These subtle but significant architectural changes can lead to measurable gains in overall efficiency. Furthermore, the use of advanced passivation techniques reduces the number of defects within the semiconductor material, minimizing recombination losses and enhancing the performance of the cell.

Another significant area of research focuses on improving the management of heat generated within PV cells. Excess heat can negatively affect cell performance and lifespan. The implementation of advanced cooling strategies, such as microfluidic cooling or heat-sink integration, can effectively dissipate heat, leading to enhanced efficiency and durability. This is especially important for high-concentration PV systems, where heat generation can be more pronounced. Furthermore, the incorporation of innovative thermal management materials can improve heat dissipation and maintain optimal operating temperatures.

Finally, we must consider the importance of holistic system design optimization. Even with highly efficient individual cells, the overall system performance can be limited by other factors such as the inverter efficiency, wiring losses, and shading effects. A comprehensive approach to system design, incorporating advanced modelling and simulation techniques, is needed to identify and mitigate these losses, maximizing the energy output of the entire solar array. This includes optimizing the angle and orientation of the panels based on local geographical conditions and maximizing the amount of sunlight received over the year. Therefore, efficiency improvements are not solely a matter of material science, but also a strategic effort across multiple engineering disciplines.

2.3 Emerging PV Materials

Perovskites, a class of materials with a unique crystal structure, are rapidly gaining traction in the photovoltaic field. Their exceptional ability to absorb sunlight, coupled with their relatively low manufacturing costs, makes them a compelling alternative to traditional silicon-based solar cells. Research consistently shows higher power conversion efficiencies compared to other thin-film technologies, promising a significant leap in solar energy harvesting. However, challenges remain, particularly concerning long-term stability and the toxicity of some perovskite components. Scientists are actively addressing these issues through material engineering and novel device architectures, exploring ways to improve their durability and minimize environmental impact, making them a highly promising area of ongoing research and development. Further exploration into cost-effective and environmentally friendly synthesis methods will be vital to their widespread adoption. Organic photovoltaic (OPV) cells offer a distinct set of advantages, including flexibility, low weight, and the potential for large-area, low-cost manufacturing using printing techniques. Their inherent flexibility allows for integration into unconventional surfaces and applications, expanding the possibilities beyond traditional rigid panels. While currently lagging behind silicon in efficiency, ongoing advancements in material design and device optimization are steadily improving their performance. The potential for mass production using roll-to-roll printing methods suggests a path towards significant cost reductions, making them a viable option for various applications, especially in flexible electronics and wearable devices. The future of OPVs lies in further boosting efficiency while maintaining their inherent advantages.

Quantum dots (QDs), nanoscale semiconductor crystals, present another exciting avenue for PV technology. Their unique size-tunable optical properties allow for precise control over light absorption, enabling the creation of multi-junction solar cells that capture a wider range of the solar spectrum. This broader absorption range theoretically leads to higher overall energy conversion efficiencies. The potential for high efficiencies and cost-effective manufacturing has driven significant research investment. However, QD-based solar cells currently face challenges related to stability and cost-effectiveness, particularly in relation to scaling up production. Overcoming these hurdles is crucial for realizing their full potential as a high-efficiency, cost-competitive solar energy solution.

Dye-sensitized solar cells (DSSCs) are known for their unique approach, utilizing a dye-sensitized titanium dioxide layer to capture light. This method allows for relatively low-cost manufacturing and offers the potential for semi-transparency, making them suitable for building-integrated photovoltaics (BIPV). DSSCs' potential for application in aesthetically pleasing and energy-efficient buildings is substantial. Their lower manufacturing temperatures and less demanding production processes could significantly lower the barrier to entry for small-scale solar cell manufacturing. Despite their advantages, efficiency remains a limiting factor compared to silicon and other emerging materials, hindering broader market penetration. Further improvements in efficiency and stability are necessary to compete effectively in the wider PV market.

Beyond these leading candidates, research continues to explore numerous other materials, including hybrid organic-inorganic perovskites and advanced silicon alloys, aiming to push the boundaries of solar cell performance. These explorations encompass diverse approaches and seek to combine the best features of various materials to achieve higher efficiencies, lower costs, and enhanced stability. The pursuit of novel material combinations and device structures fuels innovation and drives the ongoing progress in solar energy technology. The exploration into novel materials is essential to meet the ever-increasing demand for clean and sustainable energy. The development and improvement of these emerging PV materials will be key to furthering the growth and progress of the solar energy industry. Understanding the advantages, challenges, and future potential of each material is crucial for researchers, engineers, and policymakers working towards a brighter, more sustainable energy future.

3 Chapter 3: Concentrated Solar Power

Concentrated solar power (CSP) offers a compelling alternative to photovoltaic (PV) systems, particularly in regions with high solar irradiance and ample land availability. Unlike PV, which directly converts sunlight into electricity, CSP uses mirrors or lenses to concentrate sunlight onto a receiver, generating heat to drive a conventional power cycle. This heat can then be used to generate electricity, providing a valuable source of dispatchable power—a key advantage over PV, which is inherently intermittent.

The design of a CSP plant is crucial for its efficiency and cost-effectiveness. Several configurations exist, each with its own strengths and weaknesses. Parabolic trough systems, for instance, use long, curved mirrors to focus sunlight onto a central pipe containing a heat-transfer fluid. Power tower systems, on the other hand, employ hundreds or even thousands of heliostats—individually controlled mirrors—to reflect sunlight onto a central receiver atop a tower. Finally, dish Stirling systems utilize individual parabolic dishes focusing sunlight onto a Stirling engine, converting the heat directly into mechanical work. The choice of system depends on factors such as available land, solar resource, and desired power output. Careful consideration of these factors is paramount for optimal system design.

Thermal energy storage is an essential component of CSP systems. The intermittent nature of solar energy necessitates the ability to store the heat generated during the day for use at night or during periods of low solar irradiance. Several technologies are employed for this purpose, including molten salts, which have high heat capacity and can store significant amounts of thermal energy at high temperatures. Other options include sensible heat storage in materials like rocks or sand, and latent heat storage, which leverages the energy released or absorbed during phase changes. The choice of storage medium significantly impacts the cost and performance of the CSP plant. Effective thermal storage allows for extended operational periods, increasing the value and reliability of the system, enhancing its overall economic viability and minimizing reliance on fossil fuel backup power.

The economics of CSP are complex and depend on various factors, including the cost of land, materials, construction, and operation and maintenance. However, technological advancements are driving down costs, making CSP increasingly competitive with other power generation technologies, particularly in regions with abundant sunshine and suitable land. Government policies and incentives also play a significant role in shaping the economic landscape of CSP, creating a framework of support which aids implementation and ensures its adoption. Furthermore, the potential for long-term operation and reduced reliance on intermittent energy sources improves its profitability.

Environmental considerations are crucial for any energy technology, and CSP is no exception. While CSP plants have a relatively low carbon footprint compared to fossil fuel-based power plants, they do have environmental impacts, such as land use and water consumption. These impacts must be carefully assessed and mitigated through appropriate planning and design. Careful site selection and innovative technologies to minimize water usage are vital for making CSP a truly sustainable energy solution. Moreover, the potential for integrating CSP with other renewable energy sources is a strong benefit towards mitigating any negative effects. Lifecycle assessments are essential tools in evaluating the full environmental footprint of CSP plants, facilitating informed decision making and ensuring responsible environmental management.

Technological advancements are continually improving the efficiency and cost-effectiveness of CSP systems. Research efforts are focused on developing more efficient and durable mirrors, improving heat transfer fluids, developing advanced thermal storage materials, and enhancing the efficiency of power cycles. Furthermore, hybrid systems that combine CSP with other renewable energy technologies, such as PV or biomass, are being explored to enhance reliability and increase overall energy output. These developments are paving the way for wider adoption and deeper integration of CSP into existing power grids, making it a significant energy player in the future of renewable energy. Integrating CSP into smart grids will also improve electricity distribution and overall efficiency.

CSP is poised to play a significant role in the global energy transition. Its ability to provide dispatchable solar power addresses a crucial challenge associated with other renewable energy technologies, such as PV and wind. As the cost of CSP continues to decline and technological advancements further enhance its performance, its adoption is expected to grow significantly, contributing significantly to a cleaner, more sustainable energy future. The widespread adoption of CSP will offer a viable alternative to many traditional power generation sources, ultimately shaping a more responsible and environmentally conscious energy sector.

3.1 CSP System Designs

Concentrated solar power (CSP) systems offer a unique approach to harnessing the sun's energy, differing significantly from photovoltaic (PV) systems. Unlike PV cells that directly convert sunlight into electricity, CSP systems use mirrors or lenses to concentrate sunlight onto a receiver, generating heat. This heat then drives a conventional power generation cycle, typically using steam turbines to produce electricity. This indirect conversion method allows for efficient energy storage and dispatch, a critical advantage over PV, which is inherently intermittent.

The core of any CSP system lies in its design, a complex interplay of optical components, thermal systems, and power generation technology. Parabolic trough collectors, for instance, utilize long, curved mirrors to focus sunlight onto a receiver pipe containing a heat-transfer fluid. This fluid, often a molten salt, reaches high temperatures, transferring its thermal energy to generate steam. Power towers, another prominent CSP design, employ a field of heliostats – individually controlled mirrors – to reflect sunlight onto a central receiver atop a tower. This design offers high concentration ratios, enabling very high temperatures and greater efficiency. Finally, dish Stirling systems use a single large parabolic dish to focus sunlight onto a Stirling engine, a unique type of engine that converts heat directly into mechanical energy, subsequently generating electricity. Each design presents its own advantages and disadvantages in terms of cost, efficiency, and scalability.

Selecting the appropriate CSP design depends heavily on several crucial factors. Site-specific considerations, such as available land area, solar irradiance levels, and environmental factors, play a pivotal role. The desired capacity of the plant, the need for energy storage, and the local climate also influence the design selection. For example, a desert environment with high solar irradiance might be ideal for a power tower system, capable of achieving high temperatures and efficiencies. Conversely, regions with less intense sunlight might be better suited to parabolic trough systems, which require less precise sun-tracking and are generally less expensive to construct. Careful consideration of these variables is crucial for maximizing the economic and environmental benefits of the CSP plant.

Efficient thermal energy storage (TES) is paramount to the success of any CSP system. Because the sun's energy is only available during daylight hours, effective TES allows the plant to continue generating electricity even after sunset, providing a dispatchable power source that complements intermittent renewable energy sources like wind and PV. Molten salts, due to their high heat capacity and thermal stability, are a prevalent choice for TES in CSP plants. These salts are heated during the day and stored in insulated tanks, releasing their stored thermal energy at night to continue powering the turbine. The design and capacity of the TES system directly impact the plant's ability to supply electricity consistently, influencing its overall operational efficiency and economic viability. Advanced TES technologies, such as using thermal batteries or other innovative materials, are constantly under development, promising further improvements in energy storage capacity and cost-effectiveness.

The integration of CSP plants into the existing power grid presents additional design challenges and considerations. CSP plants, with their ability to dispatch power on demand, can be designed to complement other renewable energy sources and ensure grid stability. However, careful planning and coordination with grid operators are essential to ensure seamless integration and avoid potential power imbalances. Transmission lines, substations, and other grid infrastructure must be adequately sized and upgraded to accommodate the added power generation capacity. Furthermore, smart grid technologies and advanced control systems are being incorporated to manage the output of CSP plants and optimize their operation in response to varying electricity demand and solar irradiance. These technological advances are crucial for the widespread adoption of CSP and its vital contribution to a cleaner and more sustainable energy future. The ongoing innovations in CSP technology and grid management strategies ensure CSP's increasingly significant role in powering a more sustainable world.

3.2 Thermal Energy Storage

Thermal energy storage (TES) is paramount to the success of concentrated solar power (CSP) plants. Unlike photovoltaic systems, CSP plants generate electricity by using sunlight to heat a fluid, which then drives a turbine. The sun's energy isn't constantly available, however; TES systems bridge this gap, allowing CSP plants to generate electricity even when the sun isn't shining. This capability is crucial for providing a reliable and consistent power supply, mirroring the output of conventional power plants.

Several TES technologies exist, each with its own advantages and drawbacks. Sensible heat storage is the most common and straightforward method, involving heating a material (like molten salt, water, or sand) to store thermal energy. This stored heat can be later used to generate steam for the turbine, ensuring continuous operation. Molten salt is particularly popular due to its high heat capacity and relatively low cost, making it a cost-effective choice for large-scale deployments.

Latent heat storage leverages the energy required for a material to change phase—for example, from solid to liquid or liquid to gas. This phase transition often occurs at a specific temperature, making it ideal for precise temperature control. Materials like phase-change materials (PCMs) are employed, offering a significant energy density compared to sensible heat storage. However, the complexity of managing phase transitions and potential material degradation can present challenges.

Thermochemical storage exploits the heat released or absorbed during chemical reactions. This method boasts exceptionally high energy densities, but it necessitates intricate reaction cycles and complex system designs, including careful management of chemical reactions to avoid undesirable side effects. The high development costs and complexity associated with thermochemical storage currently limit its widespread application.

Choosing the optimal TES system for a CSP plant demands careful consideration of several factors. The size and scale of the plant greatly influence the system's design and capacity; a larger plant will necessitate a more substantial TES system to ensure sustained operation during periods of low solar irradiance. The climate also plays a crucial role, with systems tailored to match environmental conditions, as moisture can significantly impact the performance and longevity of the storage medium.

Moreover, the cost-effectiveness of different TES technologies varies. Molten salt storage currently offers a favorable cost-benefit ratio for large-scale deployments, while thermochemical options remain pricier due to their complexity. The efficiency of energy storage and retrieval also needs evaluation. Heat losses during storage and retrieval impact the overall efficiency of the CSP plant, so minimizing these losses through insulation and efficient heat exchangers is paramount for optimization.

Further considerations include the lifespan and maintenance requirements of the TES system. Molten salts, for example, require careful handling to prevent corrosion and leakage. PCMs may undergo degradation over time. Understanding the life cycle costs associated with different TES systems is therefore important for the financial sustainability of the CSP plant. Finally, the environmental impact of the storage materials and their production processes should not be overlooked, with careful life cycle assessments to be conducted, especially during the feasibility studies for new projects.

The ongoing research and development in TES technologies aim to improve their efficiency, reduce costs, and enhance their environmental friendliness. Advanced materials, improved heat transfer designs, and innovative system controls are among the areas actively investigated to ensure the future success and sustainability of CSP technology. As this technology continues to mature, we can anticipate more efficient and cost-effective TES systems, making CSP a more viable and competitive source of clean energy, with a significant potential to contribute towards a sustainable energy future.

The integration of advanced TES systems into CSP plants is essential for their reliable and dispatchable electricity generation. This ability to store and release thermal energy expands the operational capabilities of CSP plants, increasing their value and market appeal. Efficient TES systems address the intermittency challenge inherent in solar energy, which ultimately makes CSP plants a more valuable player in the energy market.

Ultimately, the successful deployment of CSP technology relies on advancements in TES. Improvements in storage capacity, efficiency, and cost-effectiveness are key drivers for broader adoption of this important renewable energy source. The continuous development of novel materials and system designs fuels the progress and further integration of advanced TES solutions, driving towards a clean and sustainable energy landscape. Through this ongoing progress, TES is no longer just a supporting player but a key component in the future of solar power generation, and its continued improvement ensures the success of solar energy in the years to come.

4 Chapter 4: Solar Energy Applications

Residential solar systems offer a practical pathway to energy independence for homeowners. Installing solar panels on rooftops reduces reliance on the grid, lowering electricity bills and decreasing a household's carbon footprint. System size depends on energy consumption, roof space, and sunlight availability; a professional assessment is crucial for optimal design and efficient energy production. Homeowners can choose from various panel types, inverters, and mounting systems based on budget and aesthetic preferences. Government incentives, such as tax credits and rebates, often significantly offset upfront costs, making solar energy increasingly accessible for many households. Furthermore, the increasing availability of financing options, including solar loans and power purchase agreements (PPAs), further expands access to this technology. Careful consideration should be given to factors like local building codes, homeowner's association regulations, and potential shading issues to ensure seamless installation and long-term performance. Regular maintenance, including panel cleaning and inverter inspections, helps maximize energy generation and system lifespan.

Commercial solar installations provide significant financial and environmental benefits for businesses. Investing in solar power can drastically reduce energy expenses, enhancing profitability and improving a company's sustainability profile. The scale of a commercial system depends on energy consumption, available roof or land space, and the business's specific energy needs. Companies may choose to own their system outright or opt for lease or PPA arrangements, spreading the costs over time. Designing and installing a commercial solar system requires careful consideration of factors such as load profiles, energy storage solutions, and potential grid integration challenges. The return on investment (ROI) from a commercial solar installation can vary widely, influenced by factors such as energy prices, government incentives, and maintenance costs. However, many businesses are finding that the long-term savings outweigh the initial investment, while concurrently demonstrating their commitment to environmental responsibility.

Utility-scale solar farms are playing a pivotal role in the global shift towards renewable energy. These large-scale installations generate substantial amounts of electricity, supplying energy to entire communities and reducing reliance on fossil fuels. Factors like land availability, solar irradiance, and grid infrastructure significantly influence the location and design of utility-scale solar farms. Advanced technologies, such as tracking systems that optimize panel orientation throughout the day and sophisticated inverters that improve energy conversion efficiency, are integral to maximizing energy output. The environmental impact of utility-scale solar farms, including land use and habitat disruption, needs careful consideration and mitigation strategies often involve integrating the project with land conservation efforts. Land-use conflicts are frequently addressed by incorporating measures such as pollinator-friendly vegetation and wildlife corridors, contributing to biodiversity within the project site. Furthermore, successful implementation requires detailed planning, effective grid integration, and coordinated efforts across various stakeholders, including developers, landowners, regulatory bodies, and the local communities.

Off-grid solar power systems provide a reliable energy source in remote areas or places with limited or unreliable grid access. These systems are self-sufficient, typically combining solar panels, batteries, and an inverter. The system size is tailored to meet specific energy demands, factoring in household or community consumption, available sunlight, and desired backup power capacity. Careful consideration of battery capacity and charging cycles is crucial for ensuring adequate energy storage and system longevity. Technological advancements in battery technology are crucial for improving energy storage capacity and lifespan and reducing costs. The cost-effectiveness of off-grid solar depends heavily on factors such as the initial investment, ongoing maintenance expenses, and the cost of alternative energy sources, such as diesel generators. While off-grid solar can be more expensive initially, the long-term savings on fuel costs and the environmental benefits often justify the investment.

Transportation electrification, fueled by solar power, represents a significant step towards a sustainable transportation future. Solar energy can directly power electric vehicles (EVs) through home charging systems or indirectly support the grid, enabling wider EV adoption. Integrating solar panels into charging stations for electric vehicles not only reduces the carbon footprint of transportation but also enhances the resilience and self-sufficiency of charging infrastructure. The efficiency of solar-powered EV charging depends on factors such as the size of the solar array, battery charging rates, and energy consumption patterns. As the cost of solar panels and battery storage continues to decrease, and as government incentives continue to grow, widespread adoption of solar-powered EV charging becomes increasingly cost-effective and appealing. Furthermore, the development of innovative energy storage solutions is crucial for addressing the intermittency of solar energy and ensuring reliable EV charging availability, irrespective of weather conditions or time of day. In the long run, this synergy between solar energy and electric transportation offers a compelling approach to decreasing greenhouse gas emissions and fostering a more sustainable transportation sector.

4.1 Residential Solar Systems

Residential solar systems offer a compelling pathway towards energy independence and environmental responsibility. They represent a significant step towards reducing reliance on fossil fuels and lowering carbon footprints, while simultaneously potentially decreasing electricity bills. The initial investment, while substantial, is often offset by long-term savings and potential government incentives, making them an increasingly attractive option for homeowners.

Successful implementation hinges on careful planning and system sizing. Factors to consider include roof orientation, shading, energy consumption patterns, and available space. A thorough assessment of your home's energy needs is paramount, ensuring the system's capacity aligns perfectly with your demand. This involves analyzing historical energy usage data to determine the optimal system size to meet your needs without over\item or under-production. Professional consultation from experienced solar installers is invaluable during this phase. They can assess your property and provide expert advice on system configuration and placement for optimal efficiency.

The selection of appropriate solar panels and inverters is another critical step. Panel efficiency, durability, and warranty terms are vital considerations. Similarly, inverter choices should prioritize reliability, efficiency, and compatibility with your system's components. High-efficiency panels maximize energy production, while robust inverters ensure reliable conversion of DC power from the panels to AC power for your home. The selection process should involve careful research and comparisons of available technologies and brands, weighing their performance and warranty specifications against their cost.

Installation itself is a complex process best left to qualified professionals. Proper installation ensures optimal system performance, safety, and longevity. A poorly installed system can lead to performance issues, safety hazards, and voiding of warranties. Choosing reputable installers with a proven track record and necessary certifications is crucial for a trouble-free installation process and long-term system reliability. Checking references and reading online reviews from previous clients provides valuable insight into the installer's professionalism and service quality.

Post-installation monitoring and maintenance are essential for maximizing system lifespan and efficiency. Regular inspections can identify potential issues early, preventing them from escalating into costly repairs. Many modern systems offer remote monitoring capabilities, allowing homeowners to track performance and identify any anomalies remotely. While routine maintenance is typically minimal, professional cleaning of panels and occasional inspections can significantly extend system longevity and performance. Understanding the maintenance schedule and protocols provided by the installer is critical for ensuring optimal system function over the years.

Financial aspects also require careful consideration. While upfront costs can be significant, numerous financial incentives are available to offset these expenses. These include federal tax credits, state rebates, and local utility programs designed to encourage solar energy adoption. Exploring available options and understanding the application processes for these incentives can significantly reduce the overall financial burden of solar system installation. Creating a detailed financial model factoring in these incentives, alongside projected energy savings, provides a clearer picture of the overall return on investment.

Beyond the financial benefits, incorporating a residential solar system contributes significantly to environmental sustainability. By reducing reliance on fossil fuels and lowering carbon emissions, you actively contribute to mitigating climate change. This aligns with broader societal goals of transitioning to cleaner and more sustainable energy sources. The environmental benefits extend beyond your immediate household, contributing to a wider positive impact on the environment and future generations. Moreover, home energy independence through solar provides resilience during power outages, making your home better prepared for unexpected disruptions.

The integration of residential solar systems signifies a significant step towards a more sustainable and resilient energy future. Careful planning, responsible implementation, and ongoing maintenance ensures its long-term success, promoting both individual energy independence and broader environmental stewardship. Your commitment to adopting renewable energy solutions makes you an active participant in shaping a greener future. By carefully weighing the financial aspects and actively engaging with professionals, you can unlock the numerous benefits of residential solar, creating both economic and environmental returns. The transition to solar power is a journey of sustainable growth, paving the way for a better tomorrow. The collective effort towards energy independence enhances the robustness of the overall energy grid and reduces our vulnerability to fluctuating fossil fuel prices. It's a proactive step toward responsible energy consumption, securing a more stable and reliable energy future for your family and community alike. The future of energy is bright, powered by the sun and driven by individual commitments to sustainability.

4.2 Commercial Solar Installations

Commercial solar installations present a unique set of challenges and opportunities compared to residential systems. Scale is a key differentiator; we're talking about significantly larger arrays, often involving multiple buildings or even entire campuses. This necessitates careful planning and a more sophisticated approach to system design, encompassing factors such as shading analysis, load balancing, and potential future expansion needs. The initial investment is considerably higher, demanding a comprehensive financial analysis that accounts for long-term returns, tax incentives, and potential energy savings. It's not just about slapping panels on a roof; it's a strategic investment requiring precise calculations and long-term vision.

Unlike residential setups, commercial installations often involve more complex permitting processes. Navigating local building codes, zoning regulations, and utility interconnection requirements can be a bureaucratic labyrinth. Finding experienced professionals who understand these complexities is essential, significantly reducing the risk of delays and potential conflicts. A thorough understanding of these regulations is vital, and it often requires a team, comprising engineers, permitting experts, and contractors all working in unison. This collaborative approach is crucial.

Choosing the right equipment is paramount. Commercial systems demand robust and reliable components designed to withstand harsh weather conditions and operate efficiently over extended periods. The focus shifts from individual panel performance to the overall system's energy output and longevity. This calls for a deep understanding of different inverter technologies, racking systems, and monitoring tools – all carefully selected based on the specific needs and constraints of the installation. A cost-benefit analysis, considering factors like maintenance costs and lifespan warranties, should be at the heart of this selection.

Safety is of paramount importance. Commercial sites have higher occupancy and activity levels than residential areas, increasing the potential risks related to electrical hazards. Adherence to stringent safety standards is non-negotiable; it is mandatory. Regular inspections, proper grounding, and effective safety protocols must be in place throughout the installation process and beyond, protecting both workers and building occupants. Regular training for personnel, along with clear emergency procedures, is crucial for operational safety.

Financing commercial solar projects often involves different strategies than those used for residential systems. Power Purchase Agreements (PPAs), for example, can be very attractive to businesses, eliminating upfront capital costs while providing predictable energy expenses over an extended period. Leasing arrangements and other innovative financing options should be considered to make the project financially viable. A careful financial modeling approach is imperative, projecting energy production, cost savings, and return on investment to make an informed decision.

The long-term maintenance and operation of a commercial solar system is crucial. Regular cleaning of the panels, monitoring of the system's performance, and prompt repairs are necessary to maintain efficiency and longevity. Establishing a clear maintenance schedule and having access to qualified technicians is essential to guarantee the system's performance over many years. It's also wise to consider the possibility of future upgrades or expansions, ensuring that the system can adapt to the business's evolving energy needs. Regular inspections can predict potential issues, reducing downtime.

Successful commercial solar installations are a testament to careful planning, expertise in numerous fields, and meticulous execution. The complexity involved requires a team approach, bringing together engineers, financial experts, legal counsel, and skilled installation crews. It is crucial to establish clear communication channels between all stakeholders to overcome challenges and ensure timely completion of the project.

Beyond the technical aspects, understanding the business case is fundamental. A well-defined financial model that takes into account energy costs, incentives, and potential tax benefits will demonstrate the long-term ROI and strengthen the overall project proposal. The potential for enhanced corporate social responsibility, reducing carbon footprint and attracting environmentally conscious customers, should also be considered. There's an added level of appeal and marketability that comes with the implementation of sustainable energy practices.

Commercial solar installations are more than just a technical undertaking. They are strategic investments that demonstrate a company's commitment to sustainability, cost savings, and long-term value creation. They represent a significant move toward decarbonizing the built environment, reflecting a positive impact on the environment and a responsible approach to energy consumption. It's a smart business decision, and a contribution to a greener, more sustainable future. Therefore, it's crucial to approach this endeavor with a comprehensive strategy, detailed planning, and the expertise to make it work. The benefits extend beyond the balance sheet, showcasing a commitment to corporate and environmental responsibility.

4.3 Utility-Scale Solar Farms

Utility-scale solar farms represent a significant advancement in harnessing solar energy's potential. These massive installations, often spanning hundreds of acres, consist of countless solar panels arranged in arrays, converting sunlight directly into electricity. Their sheer size allows for substantial energy generation, contributing significantly to a region's power grid. This scale provides cost advantages through economies of scale, making solar energy increasingly competitive with traditional fossil fuel sources.

The construction of a utility-scale solar farm involves meticulous planning and execution. Site selection is crucial, considering factors such as solar irradiance, land availability, proximity to transmission lines, and environmental impact. Detailed environmental impact assessments are necessary to mitigate any negative effects on local ecosystems. Engineering and design require expertise in civil engineering, electrical engineering, and environmental science to ensure efficient energy generation and minimal environmental disruption. Construction itself involves significant logistical challenges, including the transport and installation of massive amounts of equipment and the creation of extensive infrastructure.

Once operational, a utility-scale solar farm requires ongoing maintenance and monitoring. Regular cleaning of the solar panels is essential to maximize energy output, while sophisticated monitoring systems track performance and identify any potential issues. Specialized personnel are needed for both preventative maintenance and emergency repairs. The lifespan of such farms extends for decades, requiring long-term planning and commitment from developers and investors. Furthermore, innovative approaches to land management, such as integrating agriculture or grazing within the solar farm, can create mutually beneficial synergies, optimizing land use and reducing environmental impacts.

Land use remains a key consideration. While these farms require significant land areas, innovative approaches minimize their ecological footprint. The integration of solar farms with agricultural practices, known as "agrivoltaics," allows for simultaneous crop production and energy generation, maximizing land use efficiency. This strategy benefits both energy production and food security while potentially mitigating some of the concerns about land displacement. Moreover, careful siting and responsible land management can minimize the impact on wildlife habitats and biodiversity. By incorporating ecological best practices, utility-scale solar farms can become examples of sustainable land use.

The economic impact of utility-scale solar farms extends beyond job creation during construction and operation. These projects often stimulate local economies through tax revenue generation and increased property values. Furthermore, the shift away from fossil fuels reduces reliance on volatile global energy markets, enhancing energy independence and security for communities and nations. Long-term economic benefits are significant due to the low operating costs of solar farms and decreasing solar panel costs, contributing to affordable and clean electricity. The economic benefits are enhanced by government incentives like tax credits and subsidies designed to support renewable energy development. This sustained economic impact strengthens local communities and supports broader economic growth.

Grid integration presents a critical challenge and opportunity. Utility-scale solar farms require robust grid infrastructure to handle the influx of intermittent renewable energy. Smart grid technologies, advanced energy storage solutions, and sophisticated grid management strategies are vital for seamlessly integrating solar power into the electricity supply, ensuring grid stability and reliability. These technological advancements not only improve the reliability of the grid but also increase its efficiency, reducing energy waste and improving the overall sustainability of power generation. Sophisticated forecasting models, utilizing weather data and machine learning, improve the predictability of solar energy generation, assisting grid operators in balancing supply and demand.

The future of utility-scale solar farms involves continued technological advancements, driving down costs and increasing efficiency. Improvements in solar panel technology, advanced energy storage systems, and innovative grid management strategies will further enhance the viability and sustainability of these projects. Research and development are crucial for unlocking the full potential of solar energy. New materials and designs promise even higher energy conversion rates, and improved energy storage solutions will allow for greater reliability and dispatchability of solar power, addressing the intermittency challenge. As technology continues to advance, expect to see utility-scale solar farms become an even more significant player in the global energy transition.

In conclusion, utility-scale solar farms are essential components of a sustainable energy future. Careful planning, responsible land management, and technological innovation are critical to maximizing their benefits while minimizing their environmental impacts. The economic and environmental advantages, coupled with ongoing technological advancements, position these projects as a key driver of the global transition to clean and sustainable energy sources. Their large-scale impact on energy production and grid integration makes them a vital part of the transition to a cleaner, more sustainable energy future for generations to come. Understanding their development, operation, and integration into the energy grid is crucial for anyone involved in the energy industry or interested in sustainable energy solutions.

4.4 Off-Grid Solar Power

Designing an off-grid solar power system requires careful planning and consideration of several key factors. You need to assess your energy needs accurately, determining your daily and peak power consumption. This involves identifying all appliances and devices you'll power, noting their wattage and average daily usage. For example, a refrigerator might consume 150 watts constantly, while a laptop uses significantly less, intermittently. This meticulous accounting is critical; an undersized system will leave you short, while an oversized one is wasteful.

The location of your off-grid site plays a crucial role in system design. Solar irradiance, the amount of sunlight received per square meter, varies greatly depending on geographical location, altitude, and weather patterns. Sites with consistently high solar irradiance are ideal, maximizing energy production. You'll need to research local solar irradiance data; readily available online tools and resources can assist in this crucial step, providing you with the specific data needed for your area. You should also consider the angle and orientation of your solar panels to optimize sun exposure throughout the day. South-facing panels in the Northern Hemisphere are generally most effective, but local factors like shading from trees or buildings must be accounted for.

Choosing the right solar panels is essential. Monocrystalline silicon panels generally offer higher efficiency, meaning they generate more power from the same surface area. However, polycrystalline panels are often more cost-effective. The decision hinges on your budget and energy needs. The total wattage of your solar panel array needs to match or exceed your energy requirements. Remember, solar panels produce energy only when the sun shines, so energy storage is vital.

Battery capacity is directly linked to your energy consumption and the number of days you need to operate without sunlight. Deep-cycle batteries, designed for repeated discharging and recharging, are the preferred choice for off-grid systems. The battery bank's size, measured in amp-hours (Ah), needs to provide sufficient energy storage for your anticipated load during periods of low or no sunlight. You'll need to account for factors like battery depth of discharge (DOD) – the percentage of battery capacity that can be safely used before recharging. Over-discharging damages batteries, significantly shortening their lifespan. Therefore, careful sizing of the battery bank is paramount.

A charge controller regulates the flow of electricity from the solar panels to the batteries. This crucial component protects the batteries from overcharging and damage, extending their lifespan. MPPT (Maximum Power Point Tracking) charge controllers are highly recommended for their efficiency in maximizing energy harvesting from solar panels in varying light conditions. The charge controller needs to be appropriately sized to handle the current from the solar array and to match the voltage of your battery bank. Improper sizing can lead to inefficient charging or even damage to the system.

An inverter converts the direct current (DC) electricity produced by the solar panels and stored in the batteries into alternating current (AC) electricity needed by most household appliances. Pure sine wave inverters are generally preferred, as they provide a cleaner power supply suitable for sensitive electronics. The inverter's power rating must be sufficient to handle your peak power demands. Insufficient inverter capacity can lead to overloading and system failure. Choosing a correctly-sized inverter ensures reliable and safe operation of your off-grid solar system.

Wiring is crucial and must be done correctly for safety and efficiency. Use appropriately sized and insulated wires to handle the current produced by your system. Use proper grounding and fusing to protect against electrical hazards. A poorly installed wiring system poses significant fire and electrical shock risks. Thorough understanding of electrical safety principles and practices is crucial, or consulting with a qualified electrician is strongly advised.

Finally, regular maintenance is critical to maximize your system's lifespan and efficiency. Cleaning solar panels periodically removes dirt and debris that can reduce energy output. Periodically checking battery voltage, inspecting wiring, and monitoring the performance of other components help detect problems early on. Proactive maintenance reduces downtime and ensures the reliable operation of your off-grid solar power system for years to come. These steps, combined with careful planning and the right equipment, ensure a successfully functional and sustainable off-grid system.

4.5 Transportation Electrification

Electric vehicles (EVs) are rapidly transforming the transportation sector, offering a compelling pathway towards a more sustainable energy future. Their integration with solar power presents a synergistic relationship, where clean energy generation directly supports clean transportation. This is crucial, considering the transportation sector's significant contribution to greenhouse gas emissions.

The electrification of transportation relies heavily on the availability of renewable energy sources. Solar energy, with its abundant and virtually inexhaustible supply, plays a pivotal role in charging EVs and powering the grid infrastructure that supports them. This is particularly relevant in regions with high solar irradiance, where solar power can significantly reduce reliance on fossil fuels for transportation.

One significant application is in the development of solar-powered charging stations. These stations can directly utilize solar panels to charge EVs, minimizing grid dependency and lowering carbon emissions associated with electricity generation. The integration of battery storage systems in these charging stations further enhances their reliability and efficiency, ensuring a continuous power supply even during periods of low solar irradiance.

Furthermore, the widespread adoption of rooftop solar panels on homes and businesses can contribute directly to EV charging. Excess solar energy generated can be used to charge personal EVs, significantly reducing the reliance on the electricity grid and further decreasing carbon footprint. Such setups require careful consideration of energy storage and management systems to optimize energy utilization and minimize waste. Smart grid technologies play a vital role here, enabling efficient energy distribution and consumption.

Beyond individual charging, solar energy can power larger-scale initiatives like public transportation systems. Electric buses, trams, and trains can be powered by solar energy, reducing their environmental impact substantially. This requires significant investment in solar power generation capacity and grid infrastructure to accommodate the increased energy demand of these transportation systems. Careful planning and optimization of energy distribution are critical for success.

The future of transportation electrification is closely tied to technological advancements in both solar power and EV technologies. Improved solar panel efficiency and reduced costs will make solar-powered charging more economically viable. Similarly, advancements in battery technology, such as increased energy density and faster charging capabilities, will improve the overall efficiency and convenience of electric vehicles.

However, challenges remain. Grid infrastructure needs upgrading to handle the increased demand from widespread EV adoption. Effective grid management strategies are crucial to ensure stability and reliability. Policy interventions are also needed to stimulate the adoption of EVs and solar technologies, including tax incentives, subsidies, and stricter emissions regulations.

The successful transition to solar-powered transportation necessitates a holistic approach. This involves the development of efficient and affordable solar technologies, the expansion of grid infrastructure, and the implementation of supportive government policies. Overcoming these challenges will unlock the full potential of solar energy in decarbonizing the transportation sector, paving the way for a sustainable and environmentally friendly future. This requires collaborative efforts from government, industry, and individuals alike.

The economic benefits of investing in solar-powered transportation are also considerable. The reduced reliance on fossil fuels translates to significant cost savings over the long term. Moreover, the creation of new jobs in the solar and EV industries will stimulate economic growth and create new opportunities. The positive impacts of transportation electrification are therefore both environmental and economic, contributing to a more sustainable and prosperous future.

It is also worth considering the environmental aspects beyond direct emissions reductions. The manufacturing process of both solar panels and EVs has its own environmental impact. Life cycle assessments are crucial for evaluating the overall sustainability of these technologies. Continuous improvements in manufacturing processes and the utilization of recycled materials are essential to minimize environmental footprints. This integrated approach is critical for achieving truly sustainable transportation.

Research and development are ongoing in both solar energy and electric vehicle technologies. New materials, innovative designs, and improved manufacturing techniques are continually emerging. The pursuit of increased efficiency and reduced costs remains a key driver in this rapidly evolving field. Continuous monitoring and evaluation of the environmental impact of these technologies are equally crucial. This means a concerted effort towards creating a circular economy for these technologies.

Finally, the success of solar-powered transportation depends on public awareness and acceptance. Educating the public about the benefits of EVs and solar power is essential to drive adoption and foster a culture of sustainability. Incentivizing the adoption of EVs and providing reliable charging infrastructure are key factors in fostering public confidence and encouraging the widespread adoption of this critical technology for a clean energy future.

5 Chapter 5: Policy and Economics

Government incentives are crucial for accelerating solar energy adoption. Tax credits, rebates, and feed-in tariffs directly reduce the upfront cost of solar installations, making them more financially attractive to homeowners and businesses. These incentives often come with stipulations, such as requirements for using domestically produced components or installing systems of a certain size. Furthermore, policies promoting renewable portfolio standards (RPS) mandate that a certain percentage of electricity generation comes from renewable sources, stimulating demand for solar energy. Effective policy design requires a nuanced understanding of economic principles and societal priorities. This intricate dance between policy and the market needs careful consideration to ensure widespread adoption without unintended consequences. For example, some countries have found that over-reliance on subsidies can lead to market dis-

tortions and hinder long-term growth.
Cost reduction strategies are essential to ensure the competitiveness of solar energy. Economies of scale play a vital role; as the solar industry grows, manufacturing costs decrease due to increased production volume and technological advancements. This has already been observed in the sharp decrease in the price of solar panels over the past decade. Beyond manufacturing, improvements in installation techniques, financing models, and grid integration efficiency all contribute to reducing the overall cost of solar power. Research and development are equally important, as innovations in materials science and engineering continue to improve solar panel efficiency and longevity. Such continued progress lowers the Levelized Cost of Energy (LCOE), a crucial metric that helps compare the long-term cost-effectiveness of different power generation technologies. This constant drive to reduce costs makes solar power more accessible and viable as a primary energy source.
Analyzing market growth trends reveals the accelerating adoption of solar energy globally. The deployment of solar power capacity has increased exponentially in recent years, driven by declining costs, supportive government policies, and growing environmental awareness. Different regions exhibit varying rates of growth, reflecting differences in policy frameworks, resource availability, and economic conditions. Understanding these re-

gional variations is critical for targeted policy interventions and investment decisions. Technological innovations also significantly contribute to market growth, as improvements in efficiency and durability make solar energy increasingly competitive with fossil fuels. The future trajectory of market growth depends on a combination of factors, including ongoing technological advancements, economic developments, and global climate change mitigation efforts. Careful monitoring of key market indicators, such as solar energy deployment figures, electricity prices, and policy changes, is vital for informed decision-making in the solar energy sector. Furthermore, projections based on various scenarios are often used to foresee the potential future of this market and its contribution to a sustainable energy future. This forecasting often involves complex models incorporating economic, technological, and policy variables. These models help stakeholders make sound investment decisions and policymakers design effective policies.

The interplay between policy, economics, and technology is dynamic and complex. Government policies can significantly influence the direction and pace of solar energy adoption. For example, supportive policies can incentivize innovation and accelerate cost reductions, leading to faster market growth. Conversely, poorly designed policies may create distortions or stifle innovation. A thorough understanding of the economic fundamentals of solar energy is therefore essential for policymakers. Economic factors such as electricity prices, financing costs, and consumer preferences all shape the market for solar power. Technological advancements also play a significant role, as innovations in solar cell efficiency, energy storage, and grid integration continuously improve the competitiveness and affordability of solar energy. Ultimately, the success of the transition to solar-powered energy hinges upon a careful orchestration of these interwoven factors, where effective policy design plays a crucial role in guiding market forces and accelerating the widespread adoption of this clean and sustainable energy resource. This requires continuous evaluation, adaptation, and close collaboration between policymakers, researchers, and industry players. Such ongoing engagement is critical to ensuring that the regulatory framework fosters innovation and supports the transition to a sustainable energy future.

5.1 Government Incentives

Government incentives play a pivotal role in accelerating the adoption of solar energy, a crucial component of a sustainable energy future. These incentives, offered at various levels—local, regional, national, and even international—significantly influence the economic viability of solar projects. Understanding these programs is therefore essential for both individuals and organizations looking to invest in solar energy. The sheer variety of incentives available underscores their importance in making solar a competitive energy source.

Many countries offer tax credits, rebates, and other financial benefits to homeowners and businesses who install solar panels. These range from direct cash payments to deductions on income tax returns. For example, the U.S. federal Investment Tax Credit (ITC) offers a significant tax credit for solar installations, while some states offer additional state-level incentives, effectively layering benefits. These programs dramatically reduce the upfront cost of solar, making it significantly more accessible to a wider population. Understanding the specifics of these programs, however, requires careful research, as eligibility criteria and benefit levels can vary greatly. Websites of state and federal energy agencies are great places to begin your search.

Beyond tax benefits, numerous governmental entities provide grants and loans specifically for solar energy projects. These grants often target specific demographic groups or project types, such as those focused on community solar projects or installations in underserved communities. Loans may offer attractive interest rates and terms, making solar projects more financially feasible for organizations with limited capital. It's vital to search for government funding opportunities, as application processes and deadlines vary significantly; this is often done through specialized online portals. Careful planning and diligent research are essential to make the most of these opportunities.

The impact of government incentives extends beyond simply reducing the initial investment costs. They also drive innovation and technological advancements in the solar industry. Incentive programs often prioritize the adoption of advanced technologies, such as high-efficiency solar panels or energy storage systems, accelerating their development and deployment. This market stimulation creates jobs, promotes local economic growth, and encourages competition, thereby continuously improving the efficiency and affordability of solar technology. The strategic placement of incentive programs therefore influences not only the current state of the solar sector but also its trajectory.

Beyond financial incentives, many governments enact policies that simplify the permitting process for solar projects. Streamlined permitting procedures reduce delays and costs associated with approvals, allowing for faster deployment of solar infrastructure. These bureaucratic simplifications remove administrative burdens, making solar installations more efficient. The time saved can equate to considerable cost savings for developers. These regulatory changes often support community solar projects by decreasing the number of hurdles encountered when working with multiple regulatory entities.

Local governments often play a significant part in creating a supportive environment for solar energy development through zoning regulations, building codes, and net metering policies. Zoning regulations may encourage the installation of rooftop solar panels by removing or relaxing restrictions. Building codes may mandate the use of solar-ready designs in new buildings, further fostering solar integration. Net metering policies, which allow solar energy system owners to sell excess electricity back to the grid, create incentives for increased solar adoption. This exchange benefits both the homeowners and utility companies.

Furthermore, some international organizations provide funding and technical assistance for solar projects in developing countries, bolstering their ability to utilize renewable energy resources. These projects address the unique challenges of energy access and affordability in such regions. Global collaboration in this aspect promotes sustainable development and energy security on a worldwide scale. Understanding these different levels of support is vital for developing well-rounded plans for utilizing solar power. One should investigate both local and international programs as they work towards sustainability.

In conclusion, the diverse range of government incentives for solar energy—from tax credits and grants to streamlined permitting processes and net metering policies—play a critical role in the expansion of the solar energy sector. These programs reduce upfront costs, stimulate innovation, and make solar energy more accessible to various users. Understanding and strategically leveraging these incentives are key to navigating the complexities of solar energy adoption and achieving a cleaner, more sustainable energy future. It is worth noting that navigating the intricate world of these programs requires planning and attention to detail. However, the rewards – in terms of cost savings, environmental benefits and support for a more sustainable future – are significant. So take the time to conduct comprehensive research. Your investment in due diligence will undoubtedly pay off.

5.2 Cost Reduction Strategies

Economies of scale are a powerful force in reducing solar energy costs. Larger-scale projects, like utility-scale solar farms, benefit from bulk purchasing of components, streamlined installation processes, and reduced per-unit costs. This translates directly to lower electricity prices for consumers and increased profitability for investors. Furthermore, the manufacturing sector's capacity expansion, driven by increasing demand, results in optimized production lines and reduced manufacturing expenses, impacting the overall cost of solar panels and systems significantly. Think of it like buying in bulk at a grocery store – the more you buy, the less each unit costs. This principle underpins the cost reductions experienced within the solar industry as projects increase in size.

Government incentives play a crucial role in making solar energy more affordable and accessible. Tax credits, rebates, and feed-in tariffs can significantly reduce the upfront costs associated with solar panel installation. These incentives not only stimulate consumer demand but also encourage investment in solar energy infrastructure and research and development. For example, the Investment Tax Credit (ITC) in the United States, which offers a tax credit for a percentage of the cost of a solar energy system, directly reduces the financial burden on homeowners and businesses. Such initiatives are instrumental in accelerating the adoption of solar technology and fostering a more sustainable energy future. The financial assistance lowers the barrier to entry, allowing more individuals and organizations to participate in the transition to cleaner energy sources. Moreover, the predictability of government support fosters long-term investment confidence in the industry.

Technological advancements consistently drive down

the cost of solar energy. Improvements in solar panel efficiency, for instance, mean that fewer panels are needed to generate the same amount of power, reducing both material and installation costs. Advances in manufacturing techniques, such as the development of more efficient and automated production lines, also contribute to lower manufacturing expenses. Research and development efforts continuously push the boundaries of solar technology, creating innovative solutions that increase efficiency while reducing costs. Consider the evolution of silicon-based solar cells, which have seen dramatic improvements in efficiency over the past few decades, making them increasingly competitive with traditional energy sources. This constant pursuit of innovation is key to ensuring that solar energy remains a cost-effective and sustainable choice for the future.

Reducing the cost of financing solar projects is another critical strategy. Lower interest rates on loans and access to affordable financing options make it easier for individuals and businesses to invest in solar energy systems. This is particularly important for larger-scale projects, which often require significant upfront capital investment. Government-backed loan programs and innovative financial instruments, such as green bonds, can help to alleviate this financial barrier. For example, some governments offer low-interest loans specifically for residential solar panel installations. This makes the transition to solar energy significantly more accessible for those without large upfront capital. These programs not only expand the adoption of solar, but it also aids in promoting a responsible and equitable transition towards renewable energy.

Efficient project management and streamlined installation processes can also significantly reduce costs. Careful planning, coordination of resources, and optimized construction techniques minimize installation time and reduce labor costs. The use of standardized components and modular designs also streamlines the installation process, leading to efficiency gains and reduced expenditures. Companies employing advanced project management techniques, including building information modeling (BIM) and lean construction principles, are able to complete projects faster and at lower costs, making solar energy projects more economically viable. The increased efficiency translates into lower costs for both consumers and businesses, boosting the overall attractiveness of solar technology as a cost-effective solution.

Finally, recycling and reuse of solar panel components plays an increasingly important role in cost reduction. As solar panels reach the end of their useful life, recycling their valuable materials can reduce waste and lower the cost of manufacturing new panels. The development of efficient and cost-effective recycling technologies is therefore a crucial aspect of ensuring the long-term sustainability and affordability of solar energy. Implementing robust recycling programs and incentivizing the responsible disposal of solar panels will decrease environmental impact while concurrently reducing the need for new materials. A circular economy approach to solar panel management is essential for both economic and environmental reasons, ensuring that the lifecycle of solar technology remains cost-effective and environmentally sustainable.

5.3 Market Growth Analysis

Analyzing solar energy market growth requires a multifaceted approach, considering technological advancements, economic factors, and supportive policies. Growth isn't merely about increased production; it's about expanding market penetration across diverse sectors and geographical locations. This involves examining the interplay of numerous variables, necessitating a detailed understanding of each component and how they interact to drive overall market expansion.

The cost of solar photovoltaic (PV) systems has dramatically decreased over the past decade, a key driver of market expansion. This reduction, fueled by economies of scale, manufacturing improvements, and technological breakthroughs, has made solar power increasingly competitive with traditional energy sources. Furthermore, government incentives, such as tax credits, subsidies, and feed-in tariffs, have significantly stimulated demand, making solar more accessible to both residential and commercial consumers. These factors, combined, have created a positive feedback loop, accelerating market growth through increased deployment and technological refinement.

However, understanding market growth isn't simply about tracking the installed capacity of solar PV. We need to dissect regional differences. While some regions boast robust solar markets due to abundant sunlight and favorable policies, others lag due to limited grid infrastructure, unfavorable regulatory environments, or a lack of financial incentives. Analyzing these disparities reveals crucial insights into market potential and areas requiring policy interventions or technological innovation to unlock further growth. Therefore, a detailed regional breakdown is essential for accurate forecasting and targeted investment strategies.

Furthermore, focusing solely on PV technology overlooks the potential of concentrated solar power (CSP). Although CSP currently holds a smaller market share compared to PV, its potential for large-scale energy production and thermal energy storage offers a significant pathway for growth, particularly in regions with high solar irradiance and energy storage needs. Analyzing the market growth of CSP technologies requires a separate but complementary approach, identifying specific applications and evaluating technological advancements in efficiency and storage capacity. This includes considering the environmental impacts and their effect on market adoption.

Analyzing the market growth also requires projecting future trends. This involves using various forecasting models to estimate future solar energy capacity additions, considering factors like technological innovation, policy changes, and economic growth. Accurate forecasting requires sophisticated models that incorporate historical data, current market dynamics, and expert insights on future technological advancements and policy scenarios. Such predictions are crucial for both governmental energy planning and private investment decisions. Understanding the limitations of these models and their potential biases is important, allowing for a more robust assessment.

Beyond merely examining quantitative data, qualitative factors are vital for a thorough market analysis. These factors encompass public perception, consumer preferences, and investor confidence. Positive media coverage, public awareness campaigns, and successful deployment projects can significantly influence market growth by bolstering consumer and investor confidence. Conversely, negative events, such as supply chain disruptions or policy setbacks, can significantly impede market progress. Thus, understanding and incorporating these qualitative aspects into the growth analysis is crucial for a realistic assessment.

Finally, sustainable market growth in the solar energy sector requires a holistic perspective that integrates technical, economic, and environmental aspects. This includes assessing the environmental impact of solar energy production and deployment, ensuring responsible sourcing of materials, and minimizing land use conflicts. Growth should not come at the cost of environmental sustainability; it requires responsible development strategies that address the life-cycle impacts of solar energy systems and contribute to a sustainable future. Only through considering these multifaceted elements can we truly understand and foster the sustainable and responsible growth of the solar energy market.

6 Chapter 6: Environmental Impact

The environmental impact of solar energy, like any technology, is multifaceted. A comprehensive life cycle assessment (LCA) is crucial for a balanced understanding. This involves evaluating the environmental burden from raw material extraction, manufacturing processes, transportation, installation, operation, and ultimately, decommissioning and disposal. Studies consistently show that solar energy's overall environmental footprint is significantly smaller compared to fossil fuel-based energy generation, considering greenhouse gas emissions, air and water pollution, and land use. However, nuance is necessary; the manufacturing of solar panels, for instance, involves energy-intensive processes and the use of certain materials with environmental concerns. These concerns, however, are being actively addressed through technological innovation and improved recycling practices.

Minimizing the environmental impact throughout the solar energy lifecycle necessitates a holistic approach. Responsible sourcing of materials is paramount, prioritizing recycled components and minimizing the use of rare earth elements where possible. Improving manufacturing efficiency, using cleaner production methods, and optimizing energy consumption during the production phase significantly reduces the carbon footprint associated with panel creation. The development and implementation of efficient recycling programs are essential to recover valuable materials and reduce waste going to landfills, a critical step in lessening the long-term environmental burden. This circular economy approach should be integrated into the entire lifecycle, minimizing waste and maximizing resource recovery.

Land use represents another critical aspect of the environmental impact of solar energy, particularly regarding large-scale solar farms. The scale of these projects can lead to habitat loss and fragmentation, impacting biodiversity. Mitigating these impacts requires careful site selection, prioritizing previously disturbed lands or areas with minimal ecological value. Innovative approaches such as agrivoltaics, combining solar energy generation with agriculture, can significantly reduce the land-use conflict. Furthermore, integrating solar panels into existing infrastructure, like rooftops and highways, minimizes the need for extensive land allocation for new solar facilities. This careful planning and innovative solutions are essential to reduce the impact on ecosystems.

While solar energy's operational phase produces minimal direct environmental effects, indirect impacts need consideration. Water usage during manufacturing and panel cleaning warrants attention; water scarcity regions require careful management of water resources in the solar energy sector. The potential for habitat disruption around solar farms requires thorough ecological assessment and mitigation strategies prior to project commencement. The effects of these projects on local wildlife and the environment need continuous monitoring to ensure adherence to environmentally responsible practices. These ongoing monitoring programs are key to adaptive management and ensuring minimal negative environmental impact.

Beyond the direct environmental considerations, the indirect socioeconomic impacts are equally vital. The shift toward solar energy creates employment opportunities in manufacturing, installation, and maintenance, potentially offsetting job losses in other sectors. However, potential displacement of communities near large-scale solar projects needs careful consideration, with mitigation strategies implemented to ensure equitable development and avoid marginalization. The economic benefits associated with reduced reliance on fossil fuels and

improved air quality often outweigh potential negative consequences, promoting sustainable development. A comprehensive assessment should consider these interconnected aspects, ensuring a balanced evaluation of the environmental impact.
The environmental performance of solar energy is constantly improving. Technological advancements leading to higher efficiency panels require less material for the same energy output, reducing resource consumption and the overall environmental burden. Similarly, research into innovative materials and manufacturing processes aims to eliminate toxic substances and reduce water usage. These continuous improvements ensure that the environmental footprint of solar energy continues to shrink over time, driving its potential as a truly sustainable energy source. The commitment to research and development in this area is crucial to ensuring that solar energy continues to contribute positively to environmental protection and sustainability goals. Further research into the life cycle impacts, alongside continuing innovation, is essential to making solar power as environmentally benign as possible.

6.1 Life Cycle Assessment

Understanding the environmental footprint of solar energy is crucial for its sustainable deployment. Life Cycle

Assessment (LCA) provides a structured framework for this, evaluating the impacts from raw material extraction and manufacturing through operation, decommissioning, and disposal. A comprehensive LCA considers energy consumption, greenhouse gas emissions, water usage, and waste generation at each stage.

Unlike other renewable energy sources, solar panel manufacturing requires significant energy input, predominantly in the production of silicon wafers and other components. This initial energy burden impacts the overall carbon footprint, although the operational phase significantly reduces this impact over the system's lifespan. The choice of manufacturing processes and materials also influences the environmental profile.

Transportation of materials and assembled panels to installation sites contributes to the overall carbon footprint, varying significantly depending on geographical location and transport modes. Optimization through localized manufacturing and efficient logistics can mitigate this impact, reducing reliance on long-distance shipping and associated greenhouse gas emissions. Careful consideration of transportation strategies is therefore paramount.

Operational phase impacts are generally low, predominantly concerning land use and potential habitat disruption. However, the potential for accidental spills of hazardous materials during maintenance or cleaning needs consideration. Implementing strict safety protocols and regular environmental monitoring are critical aspects to minimize the risks of negative impacts during the system's operational life.

End-of-life management poses environmental challenges. Solar panels contain valuable recyclable materials, including silicon, aluminum, and silver. Recycling infrastructure development is crucial for responsible disposal and resource recovery, minimizing landfill space and associated environmental damage. Furthermore, innovative recycling technologies are continually evolving to enhance efficiency and material recovery rates.

The environmental impact of solar energy depends greatly on several factors. Location-specific assessments, accounting for regional differences in energy sources, transportation methods, and environmental regulations, are essential for achieving reliable and location-specific estimations of the environmental impact of solar power projects. The use of renewable energy sources in the manufacturing processes significantly decreases the overall environmental impact.

LCA methodologies constantly evolve. Improvements in data collection, modeling techniques, and impact assessment approaches contribute to greater accuracy and comprehensiveness in evaluating the environmental performance of solar energy technologies. Staying informed about these advancements is crucial for informed decision-making.

While the initial manufacturing phase poses challenges, the long operational life of solar panels and their ability to generate clean energy significantly outweigh these impacts. However, continuous improvements in manufacturing processes and the development of robust recycling strategies are crucial for minimizing environmental burden and maximizing the sustainability of solar energy.

Choosing materials with lower environmental impacts during manufacturing contributes significantly to the overall sustainability of solar energy projects. Research into new materials and manufacturing processes continually focuses on reducing the energy required and waste generated throughout the entire life cycle. The transition towards sustainable practices needs active participation from industry stakeholders and policymakers.

Comprehensive LCA studies, combined with careful land-use planning and sustainable waste management, can help mitigate potential environmental impacts associated with solar energy development. By carefully considering these aspects, we can ensure solar energy contributes significantly to a more sustainable energy future. The continued development and implementation of robust LCA methodologies will play an important role in furthering the responsible and sustainable growth of the solar industry. This ensures that the positive environmental impacts of solar energy far outweigh any associated challenges.

6.2 Land Use Considerations

The environmental footprint of large-scale solar energy projects is a complex issue demanding careful consideration. While solar power offers a cleaner alternative to fossil fuels, its land use requirements can impact ecosystems and habitats. Understanding these impacts is crucial for responsible development. This necessitates a balanced approach, integrating environmental protection with the urgent need for renewable energy expansion. Ignoring this balance could undermine the very sustainability solar energy aims to achieve.

Large-scale solar farms, often situated in arid or semi-arid regions, can alter local landscapes. These areas, while seemingly less biologically diverse, often support specialized plant and animal life uniquely adapted to these conditions. The conversion of these lands for solar energy generation can lead to habitat loss and fragmentation, potentially threatening these fragile ecosystems. Careful site selection, minimizing disturbance, and implementing mitigation strategies become essential to lessen these effects.

Minimizing land use impact requires innovative approaches. One promising strategy is the integration of solar farms with other land uses, like agriculture. Agrivoltaics, the practice of combining solar energy generation with agricultural activities, allows for simultaneous food production and renewable energy generation on the same land. This dual-use approach can enhance land use efficiency and reduce the environmental burden associated with solely dedicated solar farms. Further research and development in this area could significantly lessen the need for large tracts of land specifically designated for solar energy.

Another crucial aspect involves the careful consideration of local biodiversity. Before initiating a project, a comprehensive environmental impact assessment is mandatory. This assessment must include a detailed survey of existing flora and fauna, identifying any endangered or threatened species. Such an assessment helps pinpoint potential conflicts and inform the development of tailored mitigation measures, ensuring the protection of biodiversity. Furthermore, designing solar farms with biodiversity in mind, incorporating features like pollinator habitats and wildlife corridors, can help to offset potential negative impacts.

Beyond the immediate impact on land, the manufacturing and transportation of solar panels also contribute to the overall environmental footprint. The extraction of raw materials, manufacturing processes, and transportation networks involved all have environmental consequences. Therefore, a comprehensive life-cycle assessment, considering all these phases, is necessary to accurately assess the overall sustainability of solar energy projects. Reducing the environmental impact of these aspects requires innovation in materials, manufacturing techniques, and supply chain management. Using recycled materials and developing more efficient manufacturing processes can significantly lessen the overall environmental footprint.

The potential for land use conflict is not limited to ecosystems. It also extends to competing land uses. In areas with high population density or intensive agriculture, the allocation of land for solar farms can create tension between renewable energy development and other crucial land uses. Careful planning and community engagement are essential to navigate these potential conflicts, ensuring that the development of solar energy supports rather than detracts from the overall well-being of the community. Transparency and effective communication are vital in establishing trust and support for solar projects. Meaningful community involvement ensures that the benefits and drawbacks are considered and discussed openly and transparently.

In conclusion, the responsible development of solar energy requires a holistic approach that carefully considers land use impacts. Integrating innovative techniques like agrivoltaics, conducting thorough environmental assessments, and implementing biodiversity-friendly designs are crucial to minimizing negative environmental consequences. Moreover, transparent communication and community engagement are essential for navigating potential conflicts and ensuring the sustainable and responsible implementation of solar energy projects, ultimately safeguarding both energy needs and environmental health. A future powered by solar energy must be a future that also values and protects the delicate balance of our ecosystems.

7 Chapter 7: The Future of Solar

Perovskite solar cells, with their potential for exceeding silicon's efficiency and lower manufacturing costs, represent a significant leap forward. Imagine a world where roofs aren't just coverings but energy generators, seamlessly integrated into the urban landscape. This isn't science fiction; it's the promise of advancements in flexible and transparent solar technologies, allowing for solar integration into virtually any surface. Further research into tandem cells, combining different materials to capture a broader spectrum of sunlight, will likely unlock even higher efficiency levels, fundamentally altering our energy production paradigms. The next generation of solar panels will be lighter, more efficient, and aesthetically pleasing, making them more appealing to a wider range of consumers and spurring faster adoption rates.

Beyond the materials themselves, intelligent energy

management systems are poised to revolutionize solar power utilization. These systems, leveraging sophisticated algorithms and real-time data analysis, can optimize energy production, storage, and distribution, minimizing waste and maximizing efficiency. They will predict energy demands, adjust solar panel angles for optimal sunlight capture, and seamlessly integrate with smart grids, creating a truly dynamic and responsive energy network. This predictive capability goes beyond simple weather forecasts; it anticipates consumption patterns, adapting energy flow to meet real-time needs across homes, businesses, and communities.

Grid integration, often cited as a major hurdle, is rapidly evolving. The development of advanced energy storage solutions, such as high-capacity batteries and pumped hydro storage, is critical to addressing intermittency issues inherent in solar power. These storage solutions act as buffers, smoothing out the fluctuations in solar energy output, and ensuring a consistent power supply, even during periods of low sunlight or peak demand. Moreover, smart grids, with their ability to manage and distribute energy efficiently across larger networks, are crucial in accommodating the decentralized nature of solar power generation. The transition involves upgrading existing infrastructure and developing new technologies, leading to a more robust, sustainable, and intelligent energy ecosystem.

The global energy transition depends heavily on solar energy's continued growth. Meeting the Sustainable Development Goals, particularly those related to affordable and clean energy (SDG 7) and climate action (SDG 13), is intricately linked to widespread solar adoption. Solar power offers a viable pathway to decarbonize electricity generation, reducing greenhouse gas emissions and mitigating climate change impacts. Its scalability and versatility are crucial in addressing energy poverty in developing countries and fostering sustainable economic growth. Governments worldwide need to prioritize supportive policies, streamlined permitting processes, and funding for research and development to fully unlock solar's transformative potential. The integration of solar energy into broader sustainable development strategies is not merely beneficial, but is, in fact, essential for a brighter, more equitable future.

Finally, the concept of space-based solar power (SBSP) holds immense long-term potential. Imagine massive solar farms orbiting Earth, constantly bathed in sunlight, transmitting clean energy wirelessly to receiving stations on the ground. While still in its nascent stages, SBSP could provide a virtually limitless source of clean energy, overcoming geographical limitations and seasonal variations. The technological challenges are significant, including material science, energy transmission, and cost considerations. However, the sheer magnitude of the potential rewards warrants continued research and development in this revolutionary field. SBSP represents a bold vision for the future, one that could fundamentally reshape our energy landscape and usher in an era of unprecedented energy abundance. The pursuit of this ambitious goal is vital for securing a truly sustainable energy future.

7.1 Technological Advancements

Perovskite solar cells, a relatively new technology, are rapidly gaining traction. Their potential for surpassing silicon-based cells in efficiency and cost-effectiveness is significant. Research constantly pushes the boundaries of perovskite materials, exploring novel architectures and compositions to enhance stability and performance. This ongoing development represents a crucial technological advancement in solar energy conversion, offering a promising path towards even cheaper and more efficient solar power generation. We're seeing a huge leap in efficiency and affordability, potentially changing the very landscape of the solar energy market. The integration of artificial intelligence (AI) and machine learning (ML) in solar energy systems is transforming design, operation, and maintenance. AI algorithms can analyze vast datasets from various sources, including weather patterns, energy consumption, and solar panel performance. This allows for predictive maintenance, optimizing energy production, and improving the overall efficiency of solar farms and individual installations. Such advanced analytics are driving a significant shift towards smarter, more self-sufficient solar energy systems. The potential for improved resource allocation and reduced downtime is truly transformative for the industry.

Advances in energy storage technologies are inextricably linked to the widespread adoption of solar energy. The intermittency of solar power necessitates robust and efficient energy storage solutions to ensure a continuous power supply. Lithium-ion batteries remain a dominant technology, but ongoing research is exploring alternative battery chemistries and advanced energy storage systems. Flow batteries, for instance, offer scalability and longer lifespans. Solid-state batteries hold the promise of enhanced safety and higher energy density. These breakthroughs are crucial for expanding solar energy's reach into areas previously constrained by reliability concerns. The development of affordable, high-capacity, long-lasting energy storage systems will be pivotal for a future powered by solar energy.

Beyond the improvements in existing technologies, entirely new approaches are emerging. Research into solar thermal systems continues to improve, focusing on enhanced heat transfer and more efficient thermal energy storage. This is particularly relevant for concentrated solar power (CSP) plants, where efficient energy storage is vital for dispatchable power generation. Innovations like molten salt storage systems provide longer duration storage capabilities, allowing for round-the-clock power delivery even in the absence of sunlight. Furthermore, exploring innovative concepts like advanced concentrating mirrors and novel heat transfer fluids are pushing the limits of what's achievable with CSP technology.

Quantum dot solar cells represent another promising area of research. These cells leverage the unique optical properties of nanocrystals to absorb sunlight over a wider spectrum, leading to potentially higher efficiencies. The inherent tunability of quantum dots allows for the optimization of light absorption and charge transport, potentially enabling a new generation of highly efficient and cost-effective solar cells. However, challenges remain in terms of long-term stability and scalable manufacturing processes, necessitating continued research and development in this exciting field. The potential payoff, however, could redefine the entire solar energy market.

The advancement of flexible and transparent solar cells is opening up new possibilities for integrating solar energy into everyday life. These cells, often based on organic or perovskite materials, can be seamlessly incorporated into building materials, clothing, and other sur-

faces. This increased integration will help to broaden the scope of solar energy's applications, potentially leading to more widespread adoption and a significant increase in the overall global solar energy capacity. Think of self-powered buildings, wearable electronics, and even solar-powered vehicles as the next stage of implementation.
Finally, the continued development of advanced materials is crucial for enhancing the performance and longevity of solar energy systems. Research into new materials with improved light absorption properties, higher carrier mobility, and enhanced stability is paramount. This includes the exploration of novel silicon alloys, advanced thin-film materials, and nanostructured materials that can significantly improve the efficiency and reduce the manufacturing costs of solar cells. This multifaceted approach to materials science is essential for driving further innovation and widespread implementation of solar energy technologies. The journey toward a truly sustainable energy future hinges heavily on this continuous quest for better materials.

7.2 Grid Integration Challenges

Integrating vast amounts of intermittent solar energy into existing power grids presents significant challenges. The inherent variability of solar power, fluctuating with weather conditions and time of day, necessitates sophisticated grid management strategies to maintain reliable electricity supply. Without these strategies, integrating large-scale solar can lead to instability and potential blackouts. This requires investment in advanced grid infrastructure and smart grid technologies to predict, manage, and mitigate the effects of fluctuating solar generation.

Effective grid integration hinges upon accurate forecasting of solar power output. This involves using advanced weather models, satellite imagery, and real-time data from solar installations to predict energy generation hours ahead. These predictions can then be used to optimize energy dispatch from other power sources, ensuring a balanced supply-demand profile. The accuracy of these forecasts directly impacts grid stability and operational efficiency. Improving the predictive capabilities remains a key research area, exploring machine learning and artificial intelligence to enhance the precision of these estimations.

Smart grids play a crucial role in managing the intermittency of solar power. They use advanced sensors, communication networks, and control systems to monitor and regulate electricity flow in real-time. By dynamically adjusting energy distribution and balancing power sources, smart grids can accommodate the fluctuations in solar generation, preventing instability. Real-time data analysis enables swift response to changes in solar output, maintaining grid stability even during periods of peak solar generation or sudden cloud cover. However, widespread implementation requires substantial infrastructural upgrades and the development of compatible software and communication protocols.

Energy storage solutions are vital for addressing the intermittency issue. Batteries, pumped hydro, and compressed air energy storage can store excess solar energy generated during peak production periods, allowing for dispatch during periods of low solar irradiance or high demand. This provides a buffer against solar variability and maintains a reliable power supply even when the sun is not shining. The cost-effectiveness and scalability of energy storage technologies are critical factors in successful grid integration. Ongoing research focuses on developing more efficient, affordable, and sustainable energy storage solutions.

The efficient and reliable integration of solar energy also demands substantial investment in upgrading transmission and distribution infrastructure. Existing grids may lack the capacity to handle the influx of large-scale solar power. Therefore, new transmission lines, substations, and other infrastructure upgrades are often required to accommodate solar power integration. This includes strengthening existing grid connections and establishing new connections to areas with high solar resource potential. This infrastructural development requires significant capital investment and careful planning to minimize environmental impact.

Another challenge lies in effectively balancing the grid's supply and demand. When solar energy output is high, this often necessitates reducing generation from other sources such as fossil fuel plants. Managing this transition efficiently requires sophisticated forecasting and real-time grid control mechanisms. This includes balancing the various energy sources in the grid and prioritizing renewable energy sources while maintaining grid stability. Balancing renewable energy with conventional generation remains a significant challenge requiring innovative solutions and integrated grid management systems.

Lastly, regulatory frameworks and market mechanisms are essential for successful grid integration. Clear policies and regulations are needed to incentivize investment in grid modernization and the adoption of smart grid technologies. Developing effective market mechanisms for pricing and valuing renewable energy sources, particularly intermittent ones, is crucial for fostering investment and ensuring fair competition. Regulatory certainty and well-designed markets are crucial to attracting investment and promoting innovation in the field of grid integration. The ongoing evolution of electricity markets requires adapting policies and regulations to accommodate the increasing contribution of solar power.

7.3 Global Energy Transition

The global energy transition is underway, driven by the urgent need to mitigate climate change and ensure energy security. Solar energy, with its abundant resource and decreasing costs, plays a pivotal role in this transformation. However, a complete shift requires far more than simply installing more solar panels; it demands a multifaceted approach involving technological advancements, supportive policies, and careful consideration of environmental and societal impacts. We're not simply replacing fossil fuels one-for-one; we're fundamentally reshaping our energy systems.

This transition isn't merely about technological innovation; it's a societal undertaking. Consider the implications for existing energy infrastructure – the power grids designed for centralized fossil fuel generation must adapt to accommodate the decentralized and intermittent nature of solar power. Smart grids, incorporating advanced technologies for energy storage and management, become essential for integrating the diverse energy sources of the future. This includes not only solar, but also wind, hydro, and potentially geothermal and other renewable forms, creating a truly diverse and resilient energy landscape.

Furthermore, the transition necessitates a fundamental shift in energy consumption patterns. Improving energy efficiency in buildings, transportation, and industry is crucial to reducing overall energy demand, complementing the growth of renewable energy sources. This involves adopting sustainable practices in construction, promoting the use of electric vehicles, and investing in energy-efficient industrial processes. The combined effect of reduced demand and increased renewable energy supply paves the way for a cleaner, more secure, and sustainable energy future.

The economic dimensions are equally crucial. The global energy transition presents significant economic opportunities, creating jobs in manufacturing, installation, and maintenance of solar and other renewable energy technologies. However, it also poses challenges, particularly for communities reliant on fossil fuel industries. A just transition is paramount, ensuring that the economic benefits of the shift are shared equitably, with programs and policies designed to support workers and communities affected by the decline of traditional energy sectors. This requires a proactive approach, anticipating these challenges and implementing comprehensive retraining and economic diversification programs.

Beyond the economic factors, environmental considerations remain paramount. While solar energy boasts significant environmental advantages over fossil fuels, its life cycle impacts must be thoroughly assessed and mitigated. This includes minimizing the environmental footprint of material extraction, manufacturing, and disposal, as well as careful planning and management of land use for large-scale solar projects. Life cycle assessments (LCAs) provide a framework for evaluating the environmental performance of solar technologies throughout their entire lifespan. Research and development into more sustainable materials and manufacturing processes are essential to minimize environmental impacts further.

Successful global energy transition necessitates international cooperation. Sharing knowledge, technologies, and best practices across countries is vital for accelerating the deployment of renewable energy globally. This involves strengthening international collaborations to support developing countries in their transition to sustainable energy systems, providing technical assistance, and financial support. International agreements and policies play a crucial role in setting targets and creating a supportive regulatory environment for renewable energy development worldwide.

In conclusion, the global energy transition represents a monumental undertaking, requiring a coordinated effort across multiple sectors and stakeholders. It is not simply about replacing one energy source with another; rather, it demands a fundamental transformation of energy systems, consumption patterns, and economic structures. By embracing innovation, implementing supportive policies, and addressing the social and environmental considerations, the world can harness the full potential of solar and other renewable energy sources to create a sustainable and prosperous future. The journey is challenging, but the rewards—a healthier planet and a more secure energy future—are immeasurable. The commitment to a sustainable future requires continuous learning and adaptation as we navigate this transformative period.

7.4 Sustainable Development Goals

The Sustainable Development Goals (SDGs), adopted by the United Nations in 2015, provide a comprehensive framework for global progress, aiming to achieve a better and more sustainable future for all by 2030. These 17 interconnected goals address a broad range of social, economic, and environmental challenges, creating an ambitious agenda for international cooperation. Crucially, achieving many of these goals is inextricably linked to transitioning to sustainable and renewable energy sources, a process in which solar power plays a pivotal role.

Solar energy's contribution to the SDGs is multifaceted and far-reaching. Consider, for example, SDG 7, which specifically targets affordable and clean energy for all. Solar power, with its abundant availability and decreasing costs, is a key technology for achieving universal access to electricity, particularly in remote or underserved communities lacking grid infrastructure. The decentralized nature of solar installations allows for localized energy generation, empowering communities and reducing reliance on centralized, often fossil-fuel-based, power plants.

Beyond SDG 7, solar energy also directly supports other goals. SDG 1, focused on ending poverty, benefits significantly from the economic opportunities created by the solar industry, including manufacturing, installation, and maintenance jobs. Furthermore, access to affordable and reliable energy through solar power enhances productivity and economic growth, contributing to SDG 8, which promotes decent work and economic growth.

The environmental benefits of solar energy align directly with multiple SDGs. SDG 13, climate action, is profoundly influenced by the significant reduction in greenhouse gas emissions achieved through widespread solar adoption. Solar energy's inherent cleanliness contrasts sharply with the pollution associated with fossil fuels, contributing to cleaner air and water, essential aspects of SDG 3, good health and well-being, and SDG 6, clean water and sanitation. The reduced environmental impact also contributes positively to SDG 15, life on land, minimizing habitat disruption compared to other energy sources.

However, the integration of solar energy into the global energy system isn't without challenges. Land use associated with large-scale solar farms is a concern, necessitating careful planning and consideration for biodiversity preservation, aligning with SDG 15's emphasis on protecting terrestrial ecosystems. Responsible sourcing of materials used in solar panel manufacturing is paramount to ensure the industry's sustainability and its contribution to responsible consumption and production (SDG 12). Promoting equitable access to the benefits of solar energy is crucial for achieving SDG 10, reduced inequalities, ensuring that the advantages aren't concentrated in specific regions or socioeconomic groups.

Addressing these challenges requires a multi-pronged approach, encompassing technological advancements, supportive policy frameworks, and international collaboration. Technological innovations aimed at improving solar panel efficiency, reducing manufacturing costs, and developing more sustainable materials are crucial for making solar power even more accessible and environmentally benign. Government policies can incentivize solar adoption, accelerate deployment, and ensure responsible development practices. Furthermore, international cooperation is essential for sharing knowledge, best practices, and technology, fostering a global transition to clean energy solutions.

The achievement of the SDGs depends on a concerted effort by governments, industry, and individuals. Solar energy, with its vast potential, plays a critical role in this endeavor, offering a pathway towards a sustainable and prosperous future. By leveraging the power of the sun, we can significantly progress toward achieving many of the SDGs, fostering a world where clean energy is universally available, poverty is eradicated, and environmental degradation is minimized, thereby creating a more equitable and sustainable world for generations to come. Investing in solar energy isn't simply an investment in technology; it's an investment in a brighter, more sustainable future for all. It's a concrete step towards a world where progress and environmental responsibility go hand in hand, shaping a future where sustainable development is not just a goal, but a reality. The continued expansion of solar energy's contribution to sustainable development will hinge on technological improvements, responsible governance, and a sustained global commitment to renewable energy sources.

7.5 Space-Based Solar Power

Harnessing the sun's energy from space presents a unique set of challenges and opportunities. Unlike terrestrial solar farms, space-based solar power (SBSP) systems would operate continuously, unhindered by weather patterns or the diurnal cycle. This constant energy stream could revolutionize global energy grids, providing a reliable and sustainable power source. Imagine a future where energy scarcity is a distant memory, a future powered by a constellation of solar satellites. The concept is audacious, certainly, but the potential benefits are immense, promising a cleaner, more stable energy future for generations to come.

The core principle behind SBSP is straightforward: collect solar energy in space, where sunlight is far more abundant and consistent, and then transmit that power wirelessly to Earth. This involves launching massive solar arrays into geostationary orbit, an altitude where satellites remain stationary relative to the Earth's surface. These arrays would capture solar radiation using high-efficiency photovoltaic cells, converting it into electricity. The challenge, however, lies in efficiently transmitting this electricity over vast distances. Microwave or laser beams are the leading contenders for this task, posing significant engineering hurdles. However, recent advancements in materials science and wireless power transmission technologies suggest that these challenges are not insurmountable.

Developing such a system demands meticulous planning and considerable resources. Launching and assembling the vast infrastructure required in space is an immense undertaking, requiring significant investments in both rocketry and space-based construction technologies. The cost of deploying SBSP is presently a major barrier to widespread adoption. However, economies of scale and technological advancements in space transportation could potentially mitigate these costs over time. It's important to consider that the long-term benefits – reliable, clean energy on a global scale – might outweigh the substantial upfront investment. The financial model needs careful evaluation, considering long-term return on investment coupled with technological innovations.

Environmental considerations are paramount. The manufacturing process for SBSP components necessitates careful consideration of its environmental footprint. Minimizing the use of rare earth minerals and selecting environmentally friendly materials are crucial steps towards ensuring the sustainability of the entire system. Moreover, the potential environmental impact of the microwave or laser beams used for power transmission needs thorough investigation and mitigation strategies. We must ensure that the benefits of clean energy aren't offset by unforeseen ecological consequences. Therefore, detailed life cycle assessments are vital to ensure the environmental soundness of SBSP.

Aside from environmental concerns, the technical hurdles are substantial. Creating and maintaining such a large-scale space-based infrastructure is incredibly complex. The durability of the solar panels in the harsh environment of space is vital; they must withstand extreme temperature variations, radiation, and micrometeoroid impacts. The efficiency of the energy transmission system is also critical, to minimize energy losses during the journey from space to Earth. Regular maintenance and repairs of the orbiting infrastructure will also be necessary. The development and refinement of robotic systems and space-based repair techniques are crucial for the long-term viability of SBSP.

The geopolitical implications of SBSP are also worth considering. Access to this vast, clean energy source has the potential to reshape international relations. The equitable distribution of this resource among nations, and collaborative international efforts in its development and deployment, is vital for fostering global cooperation and mitigating potential conflicts. The benefits of space-based solar power must be accessible to all nations. International agreements and collaborations will be necessary to ensure that this technology serves the global community as a whole. The management of this potentially transformative technology is a concern we must address promptly and transparently.

Despite these challenges, the potential rewards of SBSP are compelling. A continuous, clean, and abundant energy source could drastically alter the global energy landscape. It could help to reduce reliance on fossil fuels, mitigate climate change, and improve energy security worldwide. Moreover, SBSP could stimulate advancements in numerous related fields, such as space technology, materials science, and wireless power transmission. The long-term vision is one of abundant, clean energy, readily accessible to all, marking a paradigm shift in our approach to power generation. The prospect of a cleaner, more sustainable future powered by the sun, from space, is one that we should actively pursue. The benefits could be transformative.

www.ingramcontent.com/pod-product-compliance
Lightning Source LLC
LaVergne TN
LVHW022006200926
848481LV00007B/2258
* 9 7 9 8 3 4 8 5 8 2 3 6 4 *